Lewis Clark's field guide to

Wild flowers of the sea coast

in the Pacific Northwest

Compiled and photographed by

Lewis J. Clark

Edited and composed by

John G. S. Trelawny, B.Sc.

Design by John Houghton

Gray's Publishing Limited, Sidney, British Columbia, Canada

1. SHORE BLUE-EYED GRASS, *Sisyrinchium littorale*

A dwarf plant with short broad leaves. Dark purplish-blue flowers have very brilliant orange-yellow centres, which give rise to a large capsule. The tips of the tepals end in an abrupt, short, firm point. The stem is wing-margined and unbranched. This is a very variable species which may represent the coastal extreme of a complex, but we have followed the classification of Calder and Taylor and of Hultén, who have recognized this as a distinct species. Found in short turf above seaside rocks, in cliff crevices, on sand beaches and river flats, and in marshes and sedge-meadows.

Introduction

Dr. Lewis J. Clark's successful work *Wild Flowers of British Columbia* created the demand for these field guides containing most of the large book's colour plates and extending the coverage to include the broader field of the Pacific Northwest. Dr. Clark had completed the first two field guides and had outlined the remaining four prior to his sudden passing. The publishers, with the encouragement of Mrs. Clark, decided to complete the series in his memory with the assistance of Mr. John G. S. Trelawny, B.Sc., Dept. of Biology, University of Victoria.

It is indeed a very great honour to be asked to continue the work of such a well known personality and prominent botanist as Dr. Lewis Clark, by completing the series of field guides which he was preparing at the time of his death. In this field guide and the two that follow, we have attempted to describe each plant illustrated by Dr. Clark's beautiful coloured photographs as he did with so much feeling in the first two. However, it is difficult to emulate the work of a person so completely absorbed by the wonders of plant life, who had such a uniquely scholarly style of writing; so we have concentrated mainly on producing what we feel is a practical working description of each species within the space available, guided by Dr. Clark's outlines in his outstanding work *Wild Flowers of British Columbia,* and his preparatory notes for this series.

Included in this field guide are most of the more conspicuous flowering plants that may be encountered along the seashore from Alaska to Northern California. Those, in fact, which inhabit the sandy beaches and dunes, the rocky salt-sprayed cliffs, the grassy coastal bluffs, or the thickly forested shores — the fascinating scene of which our coastline is comprised. A vast ocean bordering our range gives a certain mildness to the climate on the coast, which has a strong influence on the vegetation. These rather unique conditions of physical environment and climate result in a characteristic population of plant life, so that this area falls naturally into one of the six environmental zones into which we have divided the Pacific Northwest.

The visitor to the sea coast will be intrigued by the interesting communities of plants that have become adapted to a life of survival on the rugged shoreline, each one playing its part in a complex network of interdependence. On the dunes and sandy beaches, for instance, where the frequent ocean gales keep the sand grains almost continually moving, the seeds of many plants will be dispersed. Most of these germinate, but only those few that possess the critical adaptations necessary for survival in such rugged conditions will be successful. The seeds must be much larger and heavier than the sand-grains, as are those of the Gray Beach Pea, *Lathyrus littoralis,* so that they become buried and have some stability during the critical germination period. Most beach-adapted plants send down deep tap-roots to reach and store moisture, and their shoots will grow up through great depths of sand, as do those of the Beach Silver-top, *Glehnia leiocarpa.* Also to assist in anchorage, this plant has sheathing petioles (leaf stalks) on the half-buried stem, the axils of which hold quantities of sand for this purpose. The Sand Verbenas, *Abronia* ssp., overcome this question of anchorage by their thick, rubbery, horizontal stems that spring from a deep tap-root, and which sprawl widely over the sand. Their broad succulent leaves are covered with sticky hairs and these catch large amounts of sand which give weight to hold them in place. Other examples also have trailing stems such as the attractive Beach Morning-Glory, *Convolvulus soldanella,* and the colourful, more bushy Beach Pea, *Lathyrus japonicus.* All these species and others may be found playing a part in anchoring sand, thus providing stability for their mutual benefit.

A visit to the coastal shores is likely to include a river or creek estuary where acres of silt have accumulated over the centuries into flat-land that is inundated at high tide. Here mud-flats are often held together by covering mats of Glasswort, *Salicornia pacifica.* Not particularly attractive, its flowers are almost microscopic, but nevertheless it is interesting for the quality of being able to withstand complete submergence by salt water at high tide. The binding action of mats of this plant assist other species to establish themselves, a critical stage in the natural process of land reclamation.

Quite different features may be found in species which survive on the rocky spray-swept cliffs that form so much of our coastline. These are plants whose searching roots find anchorage and sufficient moisture in rock crevices; for example the attractive Mist Maidens, *Romanzoffia tracyi,* whose succulent leaves are protected from dessication by the salt spray by a heavy wax cuticle. On the other hand, the foliage of Hairy Cinquefoil, *Potentilla villosa,* a coastal species of our beautiful Cinquefoils, is silvery with long soft hairs so that it sheds the salt water before it can penetrate its pores. The showy composite-flowered Gum Weed, *Grindelia integrifolia,* that finds a foothold in the driest of rocky crags, exudes a sticky gum that helps to prevent dessication. There is also the Sea Pink or Thrift, *Armeria maritima,* whose tap-root anchors tight cushions of tough linear leaves. The round pink flower-heads bear corollas that are dry and papery — adaptations to withstand salt spray and searing winds.

Some plants thrive in a wide range of habitats. The Black Hawthorn, *Crataegus douglasii*, for instance is common on the coast in rather open places, but it is sufficiently versatile to live successfully in creek beds in the dry interior and on the lower slopes of the Rockies. Serviceberry (Saskatoon), *Amelanchier alnifolia*, has, as its alternate name implies, varieties which are even more widespread. These and some other species with widely extended ranges are included in this booklet, because their prevalence on the sea coast makes this seem most appropriate.

It is interesting to note how many plants of our Pacific Northwest sea coast, especially from amongst the 'composites', have been introduced to this area from other continents. Orange Hawkweed, *Hieracium aurantiacum,* Tansy Ragwort, *Senecio jacobaea,* Tall Blue Lettuce,*Lactuca biennis,* Wall Lettuce, *Lactuca muralis,* to name a few, have found their way here from Europe or Asia and, without the rigours of competition, climate, or soil conditions of their native habitats, they have spread unchecked to become invasive weeds in our coastal gardens, waste places, and logging slashes. Brass Buttons, *Cotula coronopifolia,* however, an introduction from South Africa, seems rather specialized, so that it has not spread widely and is only able to survive in certain locations on our coast. We have indicated in each case in the text where a species has without any doubt been introduced from another continent; so that the letters **'I'** and **'N'** to designate whether a plant is introduced or native, used in Field Guides 1 and 2, have been omitted.

We have included within the scope of this field guide, a description of the type of situation in which each species may be found, giving a very general idea of its habitat.

The range has been given by the approximate north-south, east-west limits. As a rule, no attempt has been made to give individual locations, owing to the dictates of space. 'Our area' (the Pacific Northwest), stretches from Alaska, south through British Columbia, Washington and Oregon to northern California, and east from the coast to the timberline of the Rockies in B.C., the eastern borders of the States of Washington and Oregon, and the northern portion of the California-Nevada boundary.

Dates of flowering are included by stating only the month or months in which flowering may occur. It is meaningless to be any more detailed when we are dealing with such a wide range of altitudes and degrees of latitude in most cases.

We have continued the same general format originated by Dr. Clark in the first two field guides and have stressed the common, or English names due to some traditional resistance to the exclusive use of scientific names. Concerning the order in which the plants are presented, we have continued to use the same sequence of plant families as that used in the first two field guides, with minor dislocations of this sequence dictated by the necessity of printing vertical and horizontal pictures in pairs.

No keys have been provided, since it has been found on enquiry, that few amateur botanists use them. Further, there is less need for keys in such a small book. Rather it is hoped that the reader will be able, with constant use, to obtain a set of mental images from the pictures, so that recognition in the field will be a comparatively simple procedure.

Scientific terms for various plant structures have been used only where the dictates of space have necessitated brevity. In some cases it has been possible to include a less technical term for explanation in parenthesis. A few of the technical terms used will be explained in the illustrated glossary at the end of the book.

We are greatly indebted for the taxonomy employed to Hitchcock, Cronquist, Ownbey and Thompson in their great 5-volume work *Vascular Plants of the Pacific Northwest;* and for more northern species we are obligated to definitive studies by Hultén *(Flora of Alaska and Neighboring Territories),* and to Calder and Taylor *(Flora of the Queen Charlotte Islands),* and also to the valuable monographs by Szczawinski and T. M. C. Taylor issued by the Provincial Museum, Victoria, and to Leslie L. Haskin *(Wild Flowers of the Pacific Coast).* Not included in this series are the trees or ferns (for which several excellent illustrated manuals are available) nor the horsetails, grasses, sedges or rushes (which are of interest chiefly to specialists). However, an effort has been made to include in each book one or more plants that are representative of each of the major families of flowering plants.

We are also more than grateful to the many friends of Dr. Clark who have given so much of their time to offer advice and assistance, and who have lent us slides from their own collections to complete this series. Especially we would like to mention Mrs. J. M. Woollett, in this regard, and Mr. Adolf Ceska for his helpful comments on the text, as well as Dr. Marcus Bell and Mr. Stephen Mitchell for the use of the University of Victoria's herbarium facilities.

John Trelawny
Victoria,
British Columbia

2. GEYER'S ONION,
Allium geyeri

Ovoid bulbs covered with fibrous brown scales with a heavy network of veins send up simple stalks bearing umbels of pink flowers. Some of these may be replaced by unique, small, pointed bulbils. Blooms June-July. Found in moist, sometimes rocky places along the shores and along stream-banks. Range: s. Vancouver I. coastline, Fraser River Canyon (Lytton), n.e. Wash.

3. NORTHERN RICE-ROOT,
Fritillaria camschatcensis

The bulbs are covered by what resembles a cluster of cooked rice grains. They send up a single stem nearly 2' high with usually 3 whorls of lanceolate leaves. One or two pendant flowers, dark purplish-brown with an unpleasant odour, bloom May-July. Found in open places with a fairly high water table, tide flats and river valleys. Range: coastal Alas. to Wash. – esp. Queen Charlottes, Vancouver I. and Widby I.

4. STINGING NETTLE,
Urtica dioica
Creeping rhizomes send up dense clusters of up to 3'-high stems with opposite pairs of ovate, strongly serrate leaves. Stem and leaves are covered with bristly 'stinging' hairs. Clusters of small greenish flowers are borne at the leaf-axils, May-Sept. In moist rich soil throughout our area from sea level to subalpine. Very common along the coast in open thickets.

5. FALL KNOTWEED,
Polygonum spergulariaeforme
The much-branched stems are somewhat sprawling and punctuated with slim linear leaves. In cross-section the stems are sharply angled. 1-4 small bright flowers are borne in the axils of the upper leaves. The pink sepals have a distinctive green mid-rib. Flowering June to November or later on dry, rocky slopes of Cascades, and further e. in s.e. Wash.

6 × 1.0

7 × 0.5

6. COMMON KNOTWEED,
Polygonum aviculare
A widespread and pernicious weed in wastelands, especially around settlements in our area at low levels. An annual with prostrate or erect, much-branched stems up to 3' long with bluish-green ½-1'' elliptical leaves that have a collar-like stipule at the nodes. The tiny flowers have no petals and a green calyx edged with pink; July-Sept.

7. BEACH KNOTWEED,
Polygonum paronychia
This is a low, shrubby, brown-stemmed plant (mat-like in growth) with scaly, prostrate stems up to 6'' long. Numerous firm linear leaves are often tightly rolled to conserve moisture. White to pink flowers are clustered in the upper axils; May-Sept. Grows on gravel or sand along the sea coast. Range: Vancouver I. to n. Cal.

8 × 0·3

9 × 0·4

8. GOLDEN DOCK, *Rumex maritimus*
A sturdy plant, 2-3' tall, found in damp places at low elevations throughout our area. Lanceolate leaves up to 6" long clasp the stem in a curious and characteristic sheath. Yellow clusters of flowers extend from the top downward for more than half the length of the plant; June-Sept.

9. STRAWBERRY BLITE, *Chenopodium capitatum*
An immigrant from Eurasia, this well-branched plant, 2-3' tall, has more or less triangular leaves up to 4" long. Bright crimson rounded clusters of flower masses make this species easily recognizable; June-Aug. Widespread in wet locations, cultivated soil, and river bars throughout our area, especially in northern B.C.

10. COMMON ORACHE,
Atriplex patula

This highly variable weed may reach 2'. Upper leaves are lanceolate, toothed and alternate; but larger, more triangular, and opposite, 2-3" leaves, with a few large irregular teeth pointing forward, occur below. Flowers are small, green and inconspicuous; June-Sept. Limited to saline or alkaline soils along the coast or inland, surviving in very dry locations, throughout our area.

11. SEA BLITE,
Suaeda maritima

An insignificant common annual species of salt marshes on the sea coast from Alas. to Wash., often occurring with **12**. Depressed to ascending plants with thin leafy branches 6-12" long. Smooth, linear, fleshy leaves ½-1" long are slightly cupped. Drab, yellowish-green, tiny flowers are almost hidden in the leaf axils; July-Sept.

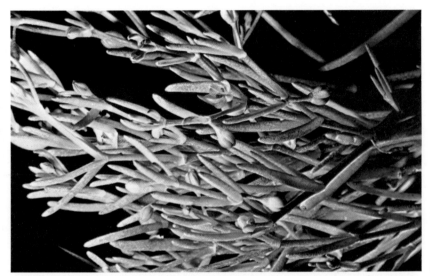

11 × 1·2

10 × 0·8

12 × 1·0

13 × 1·0

12. GLASSWORT, *Salicornia pacifica*
This very familiar plant covers thousands of acres of coastal salt marshes with its dense mats of soft, succulent, jointed, creeping stems. These are up to 3' long with 4-12" jointed, erect branches, green to reddish-purple. The tiny yellowish sunken flowers are nearly invisible; June-Sept. Common from coastal Alas. to Cal.

13. MINER'S LETTUCE, *Montia perfoliata var. glauca*
An extremely variable annual in which the terminal pair of opposite leaves are united basally. The basal leaves vary from 'thumb'-shaped to diamond-shaped with long petioles. All are succulent and edible, with tiny white to pinkish flowers blooming April to June. *Var. glauca* is extremely dwarf (1-2") of a curious greyish colour, found amongst short mosses on rocky outcrops of southern Vancouver I.

14. YELLOW SAND VERBENA, *Abronia latifolia*

This perennial spreads its prostrate rubbery stems over the sand of salt-dunes as much as 3' in many directions. Both stems and fleshy, paired, oval leaves are covered with sticky glandular hairs which hold grains of sand, helping to anchor the plant. Crowded umbels of yellow, sweet-scented flowers, less than ½" long, with flared 5-lobed mouths appear from May to Sept. Found on sandspits and sandy beaches often just above high tide level. Range: Coastal B.C. to n. Cal. Another species, PINK SAND VERBENA, *A. umbellata*, has pinkish flowers, turning to purple at the mouth of the perianth tube; and the opposite leaves are unequal. Its range is more southerly from Wash. to Cal.

15. RED MAIDS,
Calandrinia ciliata

This low, often spreading plant is much-branched from the base with sprawling succulent stems 6-9" long. The fleshy lanceolate leaves and the two unequal sepals are covered in small hairs, with petioles increasing in length as the leaves alternate downward until the basal leaves have petioles 3 times as long as the blades. Blooms, in the axils of the upper leaves, appear from April to June and last for a day. Petals, usually 5, are rose-coloured, the perianth almost ¾" across and the pistil-tip cleft into 3 stigmas. Found in gravelly soil, moist in spring time, amongst short grasses. Range: coastal s. B.C., w. Cascades Wash., e. and w. Cascades Ore.

16 × 1·0

17 × 1·1

16. PURSLANE, *Portulaca oleracea*
A pernicious annual weed introduced from Europe, is this prostrate fleshy plant that has smooth branching stems up to 12″ with obovate leaves about 1″ long. The tiny yellow flowers are found in the leaf-axils in small terminal clusters. Frequents mostly cultivated soil and waste areas at low elevations, s. B.C. to Cal.

17. SEABEACH SANDWORT, *Honkenya peploides*
A plant with sturdy stems trailing over sandy beaches, with numerous 4-10″ upright shoots. These bear mats of succulent, smooth, obovate, opposite-paired leaves recognizable by their curious yellow-green colour. In the upper leaf-axils are tucked the tiny single greenish-white flowers; May-Sept. Range: sea coast Alas. to n.w. Ore.

18. BLADDER CAMPION,
Silene cucubalus

Another introduced weed found in waste places and meadows from Alaska southward through the Pacific n.w. A perennial up to 3' tall arising from a strong taproot, much branched, with opposite obovate leaves up to 3" long united at their bases. White flowers, in compound branched inflorescences (cymes), have bell-shaped calyx tubes which later become inflated and reach over ½" in length; June-Aug.

19. SCOULER'S PINK,
Silene scouleri

Also called Scouler's Catchfly due to the sticky secretion covering the leaves and stem. An unbranched perennial about 20" tall with many opposite pubescent, oblanceolate leaves. White-petalled flowers cluster in the upper leaf-axils; the ½" calyx-tube is cylindrical; June-Aug. Common on coastal bluffs and in dry places. Range: B.C. to n. Cal., e. to Rockies.

20   × 1·5

21   × 0·4

20. PINK SAND SPURRY, *Spergularia rubra*
A common weed widespread throughout our range in dry, waste places. Its many stems, with bundles of ½", linear, dark green leaves form straggling mats up to 1' in dia. The 5 green pointed sepals, with membranous (scarious) edges, project beyond the 5 pink obovate petals. Usually 10 stamens ring the swollen green ovary, topped with its 3 styles; April-Oct.

21. SEA ROCKET, *Cakile edentula*
Possibly introduced to the Pacific shores where it is found from B.C. to California, is this sprawling, robust, fleshy-leaved smooth annual of sea- and lake-beaches. The waxed and swollen leaves are oblong to lanceolate, coarsely and irregularly toothed. The pinkish-purple flowers of July and August fade to white, and are succeeded by conspicuous oval siliques, ribbed and distended, and almost an inch long.

22. WINTER CRESS,
Barbarea orthoceras

This member of the mustard family has a single stiff and ribbed stem up to 2' tall with numerous, pinnately-divided leaves, moderately reduced upward. Similar basal leaves up to 4" long have enlarged terminal-lobes. From March to July clustered, small, bright yellow flowers, are succeeded by long straight seed pods (siliques). Frequents moist spots in woods, meadows, and waste ground from Alas. to n. Cal., e. to Rockies.

23. INDIAN MUSTARD,
Brassica juncea

This introduced annual weed has a stout 1-4' stem, branched above. The lower leaves are pinnately cleft with a large terminal lobe; the upper ones are smaller and linear. Flower spikes (racemes) up to 16" long bear pale yellow flowers, followed by upright 1"-long seed pods topped by a ¼" pointed beak; May-Aug. Found widely in established areas, in waste places, and cultivated lands from Alas. to Cal.

24 × 0.6

25 × 0.4

24. CALIFORNIA POPPY, *Eschscholtzia californica*
A deep tap-root gives rise to several 1'-tall stems with compound leaves that are much-divided. The flowers, over 2" across, have four petals, varying from pale yellow to deep orange, that open in sunlight and close at night or in cloudy weather. Vast golden masses of these flowers cover open coastal hills in Ore. and n. Cal. from May to Sept. In B.C. and Wash. it has escaped from gardens to become established locally.

25. SPREADING STONECROP, *Sedum divergens*
Horizontal rooting stems give rise to vertical 4-5" flowering branches topped with clusters of bright-yellow flowers. The thick, fleshy leaves are oval to obovate, about ¼-⅜" long, bright green to reddish. Found on rocky outcrops in full sun from sea level to high elevations, blooming from May to early Sept. according to altitude. Range: s. B.C. to Ore., coast to Cascades.

26. SEA MILKWORT,
Glaux maritima

The upright stems, about 1' tall, are crowded with opposite, fleshy, stalkless leaves. Tiny white or pinkish flowers, borne in the leaf-axils, appear from May to July. Found sporadically on coastal tide-flats, sea shores and inland saline marshes and meadowland. Range: Alas. to n. Cal.

27. BLUFF LETTUCE,
Dudleya farinosa

Smooth, glaucous plants 5-12" high in dense rosettes of thick succulent 1-2" long leaves, half as wide as their length. Stout flowering stems with broad, clasping leaves bear compact cymes (compound inflorescence) of creamy white flowers from May to Sept. Found on sea bluffs and among coastal sage or n. coastal scrub from s. Ore. to Los Angeles Co. Cal.

28. FRINGE CUP,
Lithophragma parviflora
A delicate plant of grassland and sagebrush-drylands, and common in short turf on the coast. The single flower stems, about one foot tall, bear pink to purplish flowers with deeply cleft petals from March to June. Leaves, mostly basal, are also deeply cleft. Range: s. B.C. to n. Cal., e. to Rockies.

29. RUSTY SAXIFRAGE,
Saxifraga ferruginea
Stems 6-20" tall rise from basal rosettes of thumb-shaped leaves 1-3" long. The stems are glandular-haired and strongly branching upward. From June to August appear small white irregular flowers ⅓" across, the three upper petals spotted with yellow. Leafy bublets, which fall to the ground to form new plants, are formed in the axils of the bracts. Inhabits rock crevices from coast to alpine levels throughout our range.

28 × 0.4

29 × 0.4

30. WESTERN SAXIFRAGE,
Saxifraga occidentalis
var. rufidula

This is the coastal variety of a very plastic species. *Rufidula* refers to the reddish hairiness of the lower leaf surface. A very small plant, it has 3-5 glossy, short-petioled 1-2" leaves. Usually a 2-4" scape (flower stem) carries a rather flat cluster of small white flowers with bright pink stamens. Petals are unspotted unlike other varieties, and sometimes flushed purplish, with a shallow notch. The 2 green carpels are almost completely separated, straight, and tapered to a large round pistil. Sometimes flowers in early March (near the sea) to May or later at higher altitudes. Range: B.C. s. to n.w. Ore. w. of Cascades, and up the Columbia R. Gorge to Wasco Co., Ore.

31. SERVICE-BERRY, *Amelanchier alnifolia*

A very variable shrub 5-12' tall, the reddish-brown branches become greyish with age. Alternate 1-2" smooth leaves vary from entire to strongly-toothed on the outer half, or even along the entire length. Short racemes (flower-spikes) of fragrant, showy, ¾" flowers, white, rarely flushed with pink, appear from April to July. Petals are linear to oblong, and characteristically, are slightly twisted. The 5-cleft calyx withers, but is retained at the top of the mature fruit, which is a purplish pome (like a tiny dark apple) containing a number of large hard seeds. Frequents open woodland from sea level to subalpine from s. Alas. to n. Cal., e. to Rockies.

31 × 0.8

32 × 1·0   T. & S. Armstrong

33 × 0·8

32. BLACK HAWTHORN, *Crataegus douglasii*
Shrubs or small trees with stout inch-long thorns and 1½" leaves serrate towards the tip. Flowers with 5 nearly round white petals (May-June) are succeeded by black fleshy fruits that persist into winter. Locally abundant on coastal bluffs, dry Interior creek beds, and meadowland thickets. Range: s. Alas. to n. Cal. coast, e. to Rockies.

33. TRAILING BLACKBERRY, *Rubus ursinus*
The trailing stems of this ubiquitous vine may reach 15' on coastal bluffs, forest clearings, or burned areas. This species is dioecious (male and female flowers on separate plants). The illustration shows staminate (male) flowers with no pistils, but 75-100 stamens. Their petals are longer and narrower than those of the pistillate flower which produces the blackberry up to 1" long (April-early August). Range: B.C. to n. Cal., e. to Rockies up to middle elevations.

34. THIMBLE BERRY,
Rubus parviflorus

Shrubs vary from a few in number to dense thickets over 7' high, and are usually without prickles. Leaves (to 10'' long), shaped like a maple leaf, have a soft, crinkled surface. Clusters of 3-5 white flowers, up to 2'' wide, in May to July, give way to bright red berries. Widespread from wooded, moist places to dry open areas, and from sea level to subalpine all through our range.

35. HIMALAYAN BLACKBERRY,
Rubus procerus

An escapee from Europe, with strong spreading stems up to 30' long, armed with stout flattened thorns, the compound leaves usually have 5 ovate to oblong serrate leaflets, bright green above and greyish, woolly beneath. Inflorescences of up to 20 clustered flowers with 5 white or pink ½''-long petals; June-Aug. Established on roadsides and open places from s. B.C. to n. Cal. west of Cascades, also in s.e. Wash.

36 × 0·6

36. SALMON BERRY, *Rubus spectabilis*
In damp woods these 6-12′-high thickets of arching canes armed with numerous short, straight prickles, are frequent at low levels to medium elevations, especially near the coast. Leaves are 3-foliate and doubly serrate, nearly smooth above, but pubescent beneath (at least along the veins). Showy rose-coloured blooms appear as early as February in some locations. The raspberry-like fruit (about an inch in diameter) can vary from dark purplish-red to bright orange when ripe; and is often quite sweet. Range: Alas. to n.w. Cal., e. to Cascades.

37 HABITAT

37. OCEAN SPRAY, *Holodiscus discolor*
When mature, this handsome shrub forms close clusters of strong stems 5-10' or more tall. The alternate 1½-3" leaves are lobed and toothed and hairy beneath. In June or July, it is festooned with large, cream-coloured, drooping panicles composed of hundreds of small flowers, so that it is easy to see how it came by its common name. Adorns coastal bluffs and open woodland from B.C. to n. Cal., e. to Cascades.

38. PACIFIC SILVER CINQUEFOIL, *Potentilla pacifica*
This is very similar to SILVER-WEED, *P. anserina,* except for the lower surfaces of the leaves that are only slightly silver-haired in *P. pacifica,* which also has unridged achenes (dry fruit) unlike the former. The pinnately compound leaves, sometimes over a foot long, have oblong, coarsely serrate leaflets, about ¾" long, usually green and smooth above. The flowers, ¾"-1" broad, have five bright yellow petals, usually deeper coloured towards the base and wavy-edged, appearing from May till August. This perennial plant spreads by long creeping prostrate stems that root at the nodes.

39. HAIRY CINQUEFOIL, *Potentilla villosa*

The tri-lobed, thick, long-petioled, basal leaves are deeply veined and sharply toothed; the margins beautifully outlined with silvery hairs. The whole plant appears silvered with these long, soft hairs. The flower stalks, usually having two sessile (petioleless) leaves, bear striking inch-wide golden flowers, clustered low above the foliage. The petals are sharply notched, and each bears a spot of glowing orange at its base. Found in rock crevices on coastal and montane cliffs and in arctic tundra. Range: Alas. s. to Wash., e. to Rockies. Alpine in Olympic, Cascade, and Selkirk Mountains.

40 × 1·0

40. NOOTKA ROSE, *Rosa nutkana*
This shrub, usually about 3-6' high, varies from prickly to large-thorned. The leaflets of the pinnately compound leaves are elliptical to ovate, with serrate edges. The strikingly handsome flowers may be as much as 3½" across, from light pink to deep rose. The species is characterized from May to July by a single flower at the end of each stem, and sepals with enlarged ends that are persistent. In young fruit these are carried horizontally, but later (as the roundish scarlet hips mature) vertically — in a loosely twisted crown. Common in open to open-wooded places and thickets from sea level to moderate elevations. Range: Alas. to n. Cal., e. to Selkirks in B.C., to Cascades in Wash. and Ore.

41 × 0·3

42 × 0·4

41. BROOM, *Cytisus scoparius*
A bushy shrub up to 8' tall, it has smooth-angled branches and trifoliate leaves. Originally from Europe, it was introduced on Vancouver I. in 1849 from the Sandwich Is. (Hawaii). The bright yellow flowers form carpets of gold on coastal bluffs on the islands of the Strait of Georgia and Puget Sound, and have since spread along the coast to n. Cal. and w. to the Cascades.

42. BEACH PEA, *Lathyrus japonicus*
Spreading rhizomes (rootstocks) serve to anchor colonies of this clambering plant on windswept sand dunes and beaches along the coast. The compound leaves, terminating in a long tendril, carry 6-12 smooth leathery leaflets. Halberd-shaped leaf-stipules are larger than the leaflets. Large flowers that vary from light blue to dark purple are followed by straight (up to 2½" long) pods with a short, straight beak; May-Sept. Range; coastal Alas. to n. Cal.

43 × 0·4

43. PERENNIAL PEA, *Lathyrus latifolius*

This handsome European perennial has escaped from gardens and is now widely established on roadsides, wastelands and clay banks by the sea in many areas throughout our range. The leaves consist of two large leaflets, often up to 2" broad and 6" long, with many-branched lengthy terminal tendrils by which the Pea climbs over low bushes extending 6 to 8' in every direction. Stems are angled and extended by long wing-like margins. A distinctive stipule with two unequal lobes is present at the leaf-stem junctions. The flowers, very large and showy, in colours varying from white to pale-pink or (most often) clear bright rose, bloom from May to July, and are followed by 3" long pods.

44 × 0·8 Mrs. J. Woollett

45 × 2·0

44. GREY BEACH PEA, *Lathyrus littoralis*
A covering of tiny hairs gives a greyish look to this perennial with stems up to 2′ long, and a strong rootstock anchoring it in the sand of sea shores. The compound leaves do not end in a tendril as in **43**, but in a narrow terminal leaflet. Stipules are larger than the leaflets. The flower standard is blue to purple, wings and keel pale, often white (May-July). Pods are little over 1″ long and ⅓ as wide. Range: coastal Vancouver I. to n. Cal.

45. SMALL-FLOWERED LOTUS, *Lotus micranthus*
This native annual, with stems about 1′ long, inhabits open slopes from sea level to the mountains. The 3-6 leaflets are only ¼-½″ long and nearly smooth. Flowers tiny, less than ½″ long, the pale yellow petals touched with bright rose, appear from April to Sept., followed by ¾″-long pods. Range: B.C. to n. Cal., e. to Cascades.

46. TREE LUPINE, Lupinus arboreus

A much-branched tree-like shrub reaching 5 or 6', it is native to California, but has been widely introduced northward to bind the soil on steep slopes and clay banks. It is now established in several areas from Oregon north to B.C. The unique bush-like growth makes it easy to identify among our lupines. It is silver-haired except for the upper surfaces of the leaves. Pale sulphur-yellow flowers, blooming from May to September, sometimes have a pale blue colouring.

47 × 0·9

48 × 2·0

47. SHORE LUPINE, *Lupinus littoralis*
More or less prostrate, has stems up to 2' long that trail over sandy beaches, anchored by fleshy edible roots. Leaves palmately compound with 5-8 leaflets about 1" long. Flowers (May-Aug.) in imperfect whorls on the 4" spike, have a purplish standard, often with black spots and whitish centre, and silky calyx lobes. Range: coastal B.C. to n. Cal.

48. THIMBLE CLOVER, *Trifolium microdon*
A sparsely to copiously hairy annual, seldom over 4" tall, this clover has small leaves whose leaflets are broadly obovate and deeply notched at the ends. The white to pinkish-brown flowers (Apr.-June) are subtended by a thimble-shaped, hairy whorl of fused bracts (involucre) with 10-12 irregular teeth. Found in meadows and waste ground. Range: s.w. B.C. to n. Cal., w. of Cascades.

49 × 1·0

50 × 0·3

49. COW CLOVER, *Trifolium wormskjoldii*
This tap-rooted perennial grows in coastal meadows, stream banks or salt marshes, from sea level to 8000′, blooming April to Aug. Stems up to 2′ long; the leaflets of the trifoliate leaves about 1″, oblong to obovate. The large, showy, flower-heads (over 1″ broad) are purple to pinkish, lighter at the tips. Range: B.C. to Cal. e. to Rockies.

50. GORSE, *Ulex europaeus*
As early as January and until late autumn, the showy golden flowers cluster near the branch-tips of this thick (up to 8′ tall) shrub, armed all over with stout sharp spines. Introduced from Europe, Gorse, in spite of its beauty, can be a serious weed and a fire hazard on open dry hillsides from the coast to the Cascades, s. B.C. to n. Cal.

51. STORK'S-BILL,
Erodium cicutarium

Another import of Europe, its rosettes of finely dissected leaves are widespread on dry ground throughout the lowlands of the Pacific Northwest. Almost the year round, short flower-stems arise from the leaf-axils to bear a few ¼" five-petalled magenta flowers. These are succeeded, following elongation of the pistil, by very striking 'Stork's-bills'. They consist of 5 plump carpels neatly tucked into grooves at the base of the 'bill', from each of which a very long style extends towards the tip. When the ripened fruit hits the ground, this tail twists like a corkscrew, thus acting as an auger to twist the seed into the soil, a truly marvellous adaptation.

52   × 0.6

53   × 0.6

52. HERB ROBERT, *Geranium robertianum*
This hairy, rather strong-smelling annual from Europe, has spread to certain areas of B.C., especially on the coast in the humus of open woods, on shingle, or broken rocks. It has noticeably reddish stems up to 16″ and fern-like 3-5 lobed leaves that turn bright red under dry conditions. The pretty pink ½″-wide flowers open from April onwards. Unnotched petals are often lined with white.

53. YELLOW PRAIRIE VIOLET, *Viola nuttallii*
A variable species now increasingly rare, in short turf along the coast from s.w. B.C. to n. Cal. The illustration shows this coastal variety, recognizable by its hairy leaves. A form with smoother foliage occurs east of the Cascades, B.C. to Ore. in Ponderosa Pine forests and Sagebrush flats. Yellow flowers have brownish-purple lines on the lower 3 of the 5 petals. Blooms March to May on the coast, later in the Interior.

54 × 0.6

55 × 0.5

54. BEACH SILVER-TOP, *Glehnia leiocarpa*
An extensive deep-running root-system, and widely sheathing, buried leaf-stipules serve to anchor this tough perennial on sandy beaches. The leaves are thick and leathery, waxy above and densely white-woolly beneath. White flowers (June-July) are borne in umbels on a short, strong, woolly stalk. Range: coastal Alas. to n. Cal.

55. INDIAN CONSUMPTION PLANT, *Lomatium nudicaule*
Nudicaule means naked stem, since the flowering stems that support the creamy-yellow umbels are naked, without leaves or bracts. Distinctive leaves are an unusual greyish-green and they are compound. The leaflets are oblong and entire, or slightly toothed at the apex. Blooms April to June in dry, open clay soils from sea-edge to mountain slopes. Range: s. B.C. coast to n. Cal., e. to Rockies.

56 × 1·0

57 × 1·2

56. WESTERN SNAKE ROOT, *Sanicula crassicaulis*
A strong-stemmed perennial is this common and variable species. The basal leaves are nearly round in general outline, though palmately and deeply 5-lobed. Another species, *S. graveolens*, has 3-parted and more deeply divided basal leaves. Flowers usually yellow (sometimes purplish) occur May to June, followed by fruits with hooked prickles. Range: s. B.C. to n. Cal., e. to Cascades on dry, open places.

57. SCARLET PIMPERNEL, *Anagallis arvensis*
A low, sprawling, smooth annual with square stems about 1′ long, and opposite pairs (occasionally whorls) of pointed-oval, entire and unstalked leaves. From May to July individual salmon-red, ½″-wide flowers are borne on slender stalks. Introduced from Europe, it has become established in open places on the coast in s.w. B.C., occasionally in Wash. and Ore., and in Cal.

58 × 1.0

59 × 0.5

58. PETTY SPURGE,
Euphorbia peplus

Introduced from Europe, this annual weed is common on wastelands and in gardens. The much-branched, smooth, leafy plant, up to 1' high, bears many inconspicuous yellow flowers almost hidden by the paired leafy bracts; May-Nov. Range: B.C. to n. Cal. w. of Cascades.

59. SEA PINK,
Armeria maritima

On rocky cliffs along the seacoast from Alaska to California, these neat cushions of persistent linear leaves may be found. Numerous tough flower-stems, about 6" high, are leafless, and bear hemispherical heads, ½" in diameter, of pink papery flowers from March to July. Occasionally found inland, it is also commonly called Thrift.

60 × 0.4

61 × 1.1

60. BEARBERRY, *Arctostaphylos media*
A hybrid between the prostrate KINNIKINNICK, *A. uva-ursi*, and the large shrub MANZANITA, *A. columbiana*, this is a small shrub up to 2' high with greyish foliage, paler beneath. Clusters of pure white, ¼" long, bell-like flowers, hang from branch tips in May and June, and are followed by red berries. This hybrid occurs on open, exposed slopes in coastal B.C. southward.

61. ALASKA BLUEBERRY, *Vaccinium alaskaense*
Deciduous shrubs up to 4' tall with smooth, 1¼"-long leaves, dark green above, paler below. The flowers (May-June), bronzy- to pinkish-green, bell shaped and pinched in at the mouth, occur singly in the leaf-axils on ½" stalks that are swollen just below the ovaries. Berries are bluish- to purplish-black and edible. Found in wet acid soil in forests. Range: Coastal Alas. to n.w. Ore., and e. to Cascades.

62. SHORE SHOOTING STAR, *Dodecatheon littorale*
Probably a hybrid of FEW-FLOWERED, *D. pulchellum,* and BROAD-LEAVED SHOOTING STARS, *D. hendersonii*, is this puzzling but striking species with a short (3" long) thick stem and broad erect leaves. The plant illustrated was photographed on one of the coastal promontories at Sooke, on the southern Vancouver Island coast.

63 × 0·6   T. & Y. Green

64 × 0·6

63. BEACH MORNING-GLORY, *Convolvulus soldanella*
This rather rare plant of coastal dunes has trailing stems up to about 1' long; and thick, fleshy, kidney-shaped leaves. Pinkish-purple funnel-flowers up to 2" across are clasped at the base by 2 thick bracts (modified leaves), April-Sept. Range: sea coast B.C. to n. Cal.

64. SALT-MARSH DODDER, *Cuscuta salina*
This perennial species parasitizes mainly *Salicornia pacifica* (**12**) and other host plants of salt marshes. The orange-yellow twining stem climbs anti-clockwise to the top of the host, putting out thread-like processes (haustoria) that penetrate its tissues to extract nutrients for the parasite. Clusters of small whitish flowers appear from June to Aug. Range: coastal B.C. to n. Cal.

65. SKUNK WEED,
Navarretia squarrosa

The common name of this annual is most appropriate on account of its noxious odour. The erect stem, about 1' high, has alternate leaves armed with rows of spines. It terminates in a head, about 2" in diameter, of small pale to deep blue flowers, June to Sept. Found in dry open places along the coast from s. Vancouver I. to n. Cal., e. to Cascades.

66. GROUND IVY,
Glecoma hederacea

This introduced, low, hairy perennial, 6-12" high, has long-petioled, pungently aromatic, and often purplish heart- to kidney-shaped opposite leaves. The conspicuous, tubular, purple flowers, ½-⅓" long, are borne in loose whorls at the base of the leaves, April to June. Inhabits moist roadsides and wet thickets throughout our range.

67   × 1·2

68   × 1·0

67. BABY STARS, *Linanthus bicolor*
A slender annual, it grows up to 6″ high, with opposite pairs of palmate leaves, each with 3-7 sharp-fingered leaflets. These are almost rigid and covered with short stiff hairs. From the uppermost emerge clusters of very long (½-1″) slender-tubed purplish and hairy corollas from April to June. Range: s. Vancouver I., coastal to n. Cal. on open sunny hillsides. A variety with a longer corolla extends e. to Cascades.

68. WHITE FORGET-ME-NOT, *Plagiobothrys scouleri*
A small hairy annual with prostrate to erect stems up to 8″ long. The linear leaves are without petioles, alternate above, the lower ones opposite. In April or May, usually in rather poorly drained exposed coastal headlands, can be found the fragrant, yellow-centred, 5-lobed, pure white flowers, often not more than 1″ from the ground. Range: B.C. to n. Cal., e. to Rockies (lowlands to moderate elevations).

69. MIST MAIDENS,
Romanzoffia tracyi

Tufts of glossy foliage, the blades broader than long, with 5-8 rounded lobes and petioles 2 to 3 times as long as the leaves, have a heavy, protective waxy covering. Delicate white funnel-form blossoms are in few-flowered spikes scarcely surpassing the leaves; and they have short flower-stems (peduncles) and hairy calyx lobes; March to May. Mist Maidens may be found in rock-crevices on cliffs along the open sea coast within reach of the salt spray. Range: s. Alas. to n. Cal.

70. SELF-HEAL,
Prunella vulgaris

All through our area this cosmopolitan, ubiquitous plant is found from sea level to moderate altitudes in roadside ditches and open, moist places. Square stems up to 3' tall with opposite leaves bear showy purple-to-blue flowers crowded into dense terminal spikes from June to Aug.

71. HEDGE NETTLE,
Stachys cooleyae

The single stems are strongly squared and bristly-haired, reaching 3-5' high. Oppositely-paired ovate leaves, 3-6" long, are hairy on both sides. Tubular flowers, in whorls in the axils of the upper leaves, are 1-1½" long, a magenta colour, and appear from June to Aug. Prefers swampy low ground from sea level to moderate elevations. Range: s. B.C. to s. Ore., e. to Cascades.

71 × 0.7

70 × 3.0

72. BLUE-EYED MARY,
Collinsia grandiflora
In early April to June in open woods and mossy knolls up to moderate altitudes, masses of these striking little flowers form carpets of blue. The single 2 to 10''-tall stem has opposite and roundish lower leaves, the upper ones elliptic and in whorls. The corolla tube and upper lobe are bluish-lavender (later white), the lower lobe an azure blue. Range: s. B.C. to n. Cal., e. to Cascades.

73. TOAD-FLAX,
Linaria dalmatica
This immigrant is a stout, erect perennial up to 4' tall. The many firm, pointed-ovate, clasping leaves are without leaf-stalks (petioles). 'Snapdragon-shaped' flowers have a lemon-yellow tube with a bright orange pouch on the lower lip. It is sparsely established in disturbed sandy soil throughout our range from B.C. southwards, blooming throughout the summer months.

74 × 1·2

75 × 0·6

74. DWARF ORTHOCARPUS, *Orthocarpus pusillus*
This tiny plant, 1½ to 7″ tall, can be spotted by the dark-purplish or wine colour developed by most of the leaves. The minute flowers are almost too small to see — dark chocolate with a touch of yellow from the anthers. It is believed they are pollinated chiefly by ants. Found in moist places at low elevations, April to June. Range: s. B.C. to n. Cal., e. to Cascades.

75. CLUSTERED CANCER-ROOT, *Orobanche grayana*
This is a parasite on the roots of various members of the Compositae family from which it obtains nourishment, so that green leaves (being unnecessary) are lacking. The whitish stems become brownish above and bear dense spikes of ragged brownish or purplish flowers that are tubular and 5-lobed with woolly anthers (June to Sept.). Range: s. B.C. to n. Cal., e. to Cascades.

76. CLEAVERS,
Galium aparine

A common introduced annual occurring all through our range in a variety of habitats, its stems, sometimes nearly 4' long, are weak, but are supported by other vegetation. In this they are helped by rows of tiny hooks along the angles of the square stems. From the axils of whorled linear leaves, the flower stalks bear minute white flowers from April to June, followed by globular burs. THREE-PETALLED BEDSTRAW, *G. trifidum*, within the same range, but extending to higher elevations, has much blunter leaves in whorls of 4, and the fruit is smooth.

77. SEA BLUSH,
Plectritis congesta

In rather open wooded areas and rock ledges above the sea, often with **74**, this little annual may be found. On stems from 1" to 1' high are opposite pairs of smooth oblong leaves without petioles. The densely congested rounded flower-heads are terminal, and are made up of many tiny pink tubular flowers, blooming from May to June. Range: s. Vancouver I. to n. Cal., e. to Cascades.

78 × 0·3

78. RED-FRUITED ELDER, *Sambucus racemosa*
An impressive shrub, which may reach 15 feet, it has fast-growing, pith-filled, hollow canes that are amongst the first of the coastal shrubs to unfold their very large, pinnate leaves. Each of these has a terminal leaflet and 2-3 pairs of opposite, sharp-pointed and sawtooth-edged leaflets. In April along the south coast, and onward to July, inland to the Cascades, the air is filled with the rather heavy fragrance of the big creamy-white, rounded bloom-clusters, followed by bright scarlet berries. Common on the coast in moist areas on the edges of woodlands and roadsides. Range: coastal Alas. to n. Cal. e. to Cascades.

79 × 0·3

80 × 0·7

79. SEA PLANTAIN, *Plantago maritima ssp. juncoides*
The species is adapted to saline conditions and is widespread on the fringes of salt water throughout the Northern hemisphere. This subspecies is American and is common on our coast, often on flats submerged by high tides. The fleshy, linear 12″ leaves are all basal. The flower-stems, not much longer, bear a 'head' of tiny flowers with conspicuous extended anthers from June to Aug.

80. WHITE TOP, *Aster curtus*
On dry rocky hillsides on s. Vancouver I., the Puget trough and in s.w. Oregon, these small perennial herbs send up individual leafy stems from creeping rootstocks. They are about 1″ high with many stalkless oblanceolate leaves up to 1½″ long, but shorter upwards. A compact cluster of pale yellow flower-heads terminates each stem (July-Aug.).

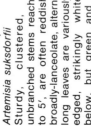

81. MILFOIL,
Achillea millefolium

Often called Yarrow, this aromatic variable perennial has been introduced from Europe and frequents dry waste places from s. Alas. all through our range. The leaves are much dissected and have a fern-like appearance. Flat-topped clusters of small, white (sometimes pinkish) composite flowers may be found from April to Oct. or later.

82. WORMWOOD,
Artemisia suksdorfii

Sturdy, clustered, erect, unbranched stems reaching up to 5′, are often reddish. The broadly-lanceolate, alternate, 5″ long leaves are variously saw-edged, strikingly white-hairy below, but green and nearly smooth above. Very small flower-heads occur chiefly in a large terminal cluster from June to Aug. Found in clay pockets on rocky shores or in upper sandy beaches. Range: coastal s. B.C. to n. Cal., sometimes e. to Cascades.

83. DOUGLAS ASTER,
Aster subspicatus

This highly variable species of purple Aster commonly gladdens sea-shores, stream banks, roadsides and open woods from July to Oct. Strong stems up to 4' high bear many lanceolate, alternately arranged leaves with short petioles that disappear upward. Flower-heads, about 1" across, have purple to rosy-purple rays, and yellowish to reddish disk flowers. Range: coastal Alas. to n. Cal., e. to Cascades, rarely at low elevations e. of Cascades.

84. TALL BLUE LETTUCE,
Lactuca biennis

In moist roadside ditches and waste ground, this immigrant from Eurasia raises its smooth tree-like stem up to 12' high in its second year. The notched and deeply-lobed leaves may be 16 to 18" long, and they exude a milky juice when broken. Very many dull-bluish flower-heads on slender stalks appear in succession from July to Sept. Range: s. Alas. to n. Cal., e. to Rockies.

Mrs. J. Woollett

85. NORTHERN APARGIDIUM,
Apargidium boreale

The basal rosette of grass-like leaves gives rise to a leafless stem which bears a solitary yellow composite flower. This attractive little perennial graces sphagnum bogs and wet meadows from the coast to moderate elevations, June-Aug. Range: coastal Alas. to n. Cal., w. to Cascades.

86. BULL THISTLE,
Cirsium vulgare

This very common weed has been introduced from Eurasia and is now widespread throughout the Pacific Northwest, along roadsides and in waste places. The prickly leaves have spines on the upper surfaces as well as the edges, and they clasp the 2-5' stem with long spiny leaf-bases. The blooms, about 1½" across, are purple to magenta and appear from July to Sept. of the second year (the plants being biennial).

87. BRASS BUTTONS,
Cotula coronopifolia

An immigrant from South Africa that has become established along our coastline on tidal flats and other moist places, this is a smooth succulent perennial up to about 10″ tall that has sprawling branches often rooting at the nodes, and variously toothed or entire lanceolate leaves that sheath the stem. The ½″-round, rayless flower-heads (Brass Buttons) are aromatic (June-Sept.). Range: coastal, B.C. to n. Cal.

88. SAND-BUR,
Franseria chamissonis

A big, sprawling perennial plant that anchors itself in the sand of sea beaches with a number of buried branches, it protects its leaves and stems with a silvery coat of tight-pressed hairs. Flowers, small and drab though numerous, occur in a tight terminal spike and in small clusters in the axils of the upper leaves. They appear from June to Sept. followed by large brown burs. Range: coastal, s. B.C. to n. Cal.

89. GUM WEED,
Grindelia integrifolia

This very common, extremely variable plant is found along the coast from Alas. s. to n. Cal. The maritime salt-marsh form (var. *macrophylla*) has impressively large flower-heads, up to 2½" across. Away from the sea spray the rays are usually broader and shorter. The immature flower-heads appear milky and gummy. Later they develop into bright yellow ray-florets and disk-florets, blooming from June till November.

90. ORANGE HAWKWEED,
Hieracium aurantiacum

A European weed established in s. Alas. and points on the B.C. and Wash. coast, e. to the Cascades, this showy plant is sometimes known in England as 'Fox and Cubs', a name nicely descriptive of the clustered small plants arising from abundant runners, and of the densely furred foliage. The orange-red colour of the ray-flowers is distinctive, so no further description is necessary. Found in waste places and pastures (June-Aug.).

91 × 1·0

92 × 1·0   Mrs. J. Woollett

91. GOLDSTARS, *Crocidium multicaule*
A fragile little yellow daisy, 2-5″ tall, that rises from a rosette of small fleshy leaves with ½-1″ petioles and spoon-shaped spatulate blades, its flowers are solitary, ½-1″ across with bright yellow ray-florets enclosing a mass of disk-flowers appearing from March to May. Found in dry, open, sandy places and cliff-ledges. Range: coastal from Vancouver I. to n. Cal., e. up Columbia R. to Portland, Ore.

92. WALL LETTUCE, *Lactuca muralis*
Yet another introduced weed from Europe, this 2′-high slender plant bears many small yellow composite flowers from July to Sept. Well established in some coastal areas from s. Vancouver I. to Ore. in moist places and roadside ditches.

93 × 1·7

94 × 1·1

93. WESTERN TANSY, *Tanacetum douglasii*
This 8 to 24″-tall perennial rises from stout rootstocks on many coastal sand dunes. It has distinctly hairy leaves which are very finely pinnate, almost like those of **81**. The flat-topped inflorescence bears up to 20 ⅜ to ⅝″-wide yellow flower-heads, each with a ring of short-strapped ray-florets; June to Sept. Range: B.C. to n. Cal. along the coast.

94. PINEAPPLE WEED, *Matricaria matricarioides*
This lowly weed, widespread in waste and especially trodden places throughout our area, is instantly recognizable by the 'pineapple' scent of its crushed flowers (May-Sept.). A small annual, it has much-branched smooth and sprawling stems, and numerous fern-like, finely-cut leaves. Flowerheads are conical and rayless. Involucral bracts are broadly oval, with chaffy margins, and the pappus is reduced to an insignificant crown of tiny scales.

95. TANSY RAGWORT,
Senecio jacobaea

Another pernicious immigrant from Europe that is poisonous to cattle, now increasing rapidly in waste ground and on roadsides, it is a handsome weed, up to 3' tall, with showy yellow flowers and smooth foliage. The terminal segments of the deeply lobed and toothed leaves are blunt-rounded. The flower-heads, having a ring of conspicuous golden ray-flowers, appear from July to Sept. Range: Wash. and Ore. coast, e. to Cascades.

96. SOW THISTLE,
Sonchus arvensis

A troublesome perennial weed of roadsides and pastures that has invaded our area from Eurasia and Africa, the 2-6' stem is clasped by smooth, stalkless, deeply lobed leaves with prickly margins. Large golden-yellow flower-heads up to 2" across, appearing July to Oct., have many strap-shaped florets, each followed by a parachute seed — the pappus attached directly to the top of the seed.

Index

Abronia latifolia 14
   umbellata 14
Achillea millefolium 81
Alaska Blueberry 61
Allium geyeri 2
Amelanchier alnifolia 31
Anagallis arvensis 57
Apargidium boreale 85
Arctostaphylos media 60
   columbiana 60
   uva-ursi 60
Armeria maritima 59
Artemisia suksdorfii 82
Aster curtus 80
   subspicatus 83
Atriplex patula 10

Baby Stars 67
Barbarea orthoceras 22
Beach Knotweed 7
   Morning-glory 63
   Pea 42
   Silver-top 54
Bearberry 60
Black Hawthorn 32
Bladder Campion 18
Blue-eyed Mary 72
Bluff Lettuce 27
Brass Buttons 87
Brassica juncea 23
Broad-leaved Shooting
   Star 62
Broom 41
Bull Thistle 86

Cakile edentula 21
Calandrinia ciliata 15
California Poppy 24
Chenopodium capitatum 9
Cirsium vulgare 86
Cleavers 76
Clustered Cancer-root 75
Collinsia grandiflora 72
Common Knotweed 6
   Orache 10
Convolvulus soldanella 63
Cotula coronopifolia 87
Cow Clover 49

Crataegus douglasii 32
Crocidium multicaule 91
Cuscuta salina 64
Cytisus scoparius 41

Dodecatheon littorale 62
   hendersonii 62
   pulchellum 62
Douglas Aster 83
Dudleya farinosa 27
Dwarf Orthocarpus 74

Erodium cicutarium 51
Eschscholtzia
   californica 24
Euphorbia peplus 58

Fall knotweed 5
Few-flowered Shooting
   Star 62
Franseria chamissonis 88
Fringe Cup 28
Fritillaria camschatcensis 3

Galium aparine 76
   trifidum 76
Geranium robertianum 52
Geyer's Onion 2
Glasswort 12
Glaux maritima 26
Glecoma hederacea 66
Glehnia leiocarpa 54
Golden Dock 8
Goldstars 91
Gorse 50
Grey Beach Pea 44
Grindelia integrifolia 89
Ground Ivy 66
Gum Weed 89

Hairy Cinquefoil 39
Hedge Nettle 71
Herb Robert 52
Hieracium aurantiacum 90
Himalayan Blackberry 35
Holodiscus discolor 37
Honkenya peploides 17

Indian Consumption
   Plant 55
   Mustard 23

Kinnikinnick 60

Lactuca biennis 84
   muralis 92
Lathyrus japonicus 42
   latifolius 43
   littoralis 44
Linanthus bicolor 67
Linaria dalmatica 73
Lithophragma parviflora 28
Lomatium nudicaule 55
Lotus micranthus 45
Lupinus arboreus 46
   littoralis 47

Manzanita 60
Matricaria
   matricarioides 94
Milfoil 81
Miner's Lettuce 13
Mist Maidens 69
Montia perfoliata
   var. glauca 13

Navarretia squarrosa 65
Nootka Rose 40
Northern Apargidium 85
   Rice-root 3

Ocean Spray 37
Orange Hawkweed 90
Orobanche grayana 75
Orthocarpus pusillus 74

Pacific Silver
   Cinquefoil 38
Perennial Pea 43
Petty Spurge 58
Pineapple Weed 94
Pink Sand Spurry 20
   Sand Verbena 14
Plagiobothrys scouleri 68
Plantago maritima
   ssp. juncoides 79

Plectritis congesta 77
Polygonum aviculare 6
   paronychia 7
   spergulariaeforme 5
Portulaca oleracea 16
Potentilla anserina 38
   pacifica 38
   villosa 39
Prunella vulgaris 70
Purslane 16

Red Maids 15
Red-fruited Elder 78
Romanzoffia tracyi 69
Rosa nutkana 40
Rubus parviflorus 34
   procerus 35
   spectabilis 36
   ursinus 33
Rumex maritimus 8
Rusty Saxifrage 29

Salicornia pacifica 12
Salmon Berry 36
Salt-marsh Dodder 64
Sambucus racemosa 78
Sanicula crassicaulis 56
   graveolens 56
Sand-bur 88
Saxifraga ferruginea 29
   occidentalis var.
      rufidula 30

Scarlet Pimpernel 57
Scouler's Pink 19
Sea Blite 11
   Blush 77
   Milkwort 26
   Pink 59
   Plantain 79
   Rocket 21
Seabeach Sandwort 17
Sedum divergens 25
Self-heal 70
Senecio jacobaea 95
Service-berry 31
Shore Blue-eyed Grass 1
   Lupine 47
   Shooting Star 62
Silene cucubalus 18
   scouleri 19
Silver-weed 38
Sisyrinchium littorale 1
Skunk Weed 65
Small-flowered Lotus 45
Sonchus arvensis 96
Sow Thistle 96
Spergularia rubra 20
Spreading Stonecrop 25
Stachys cooleyae 71
Stinging Nettle 4
Stork's-bill 51
Strawberry Blite 9
Suaeda maritima 11

Tall Blue Lettuce 84
Tanacetum douglasii 93
Tansy Ragwort 95
Thimble Berry 34
   Clover 48
Three-petalled
   Bedstraw 76
Toad-flax 73
Trailing Blackberry 33
Tree Lupine 46
Trifolium microdon 48
   wormskjoldii 49

Ulex europaeus 50
Urtica dioica 4

Vaccinium alaskaense 61
Viola nuttallii 53

Wall Lettuce 92
Western Saxifrage 30
   Snakeroot 56
   Tansy 93
White Forget-me-not 68
   Top 80
Winter Cress 22
Wormwood 82

Yellow Prairie Violet 53
   Sand Verbena 14

Glossary

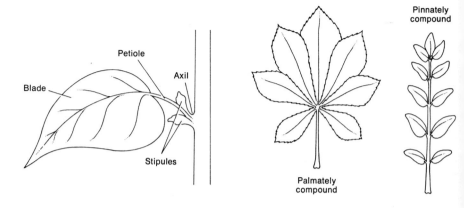

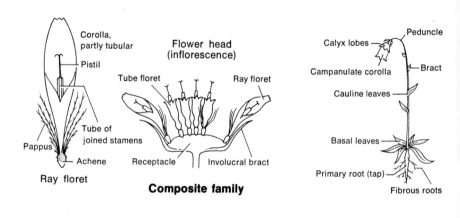

Figures in blue under each picture in the book indicate the scale of the reproduction, e.g. ×0·5 means the picture is half as large as the actual (average) plant; ×2·0 means the picture is twice as large as the plant.

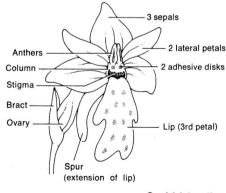